by Elizabeth Thomas

illustrated by Vicki Jo Redenbaugh

Carolrhoda Books, Inc. / Minneapolis

Gramma was very strict. She said she was
"correct."
    "There are certain proper ways that one has to
do things," she'd say.
    Everyone usually obeyed her.

"Don't just snatch some grapes willy-nilly like
that," she'd tell me. "Use the grape snips."

I'd have to use the little silver scissors with the grape clusters on the handles and carefully cut off a small, neat bunch.

"Much tidier, Dorothea," she'd say.

Every Saturday after lunch, Gramma would close
her door, recline on her chaise lounge, and listen
to the Metropolitan Opera sing on the radio,
just for her.

Sometimes she'd make me listen.
"It's by Verdi," she'd sigh. "Isn't it *thrilling?*"

She also loved her garden. She planted it with
pink roses, gold zinnias, yellow snapdragons,
purple pansies, and green beans.

Gramma had loved green beans ever since she'd had a bad stomachache and green beans were the only thing that tasted good to her.

"They're good for whatever ails you," she would say.

I preferred frozen peas.
They were prettier.

Every day Gramma would be out in her garden.
She would weed, water, and make sure that all
the plants were growing properly.

"There will be no lollygagging," she told them.
Most of the plants were obedient.

The beans were another story.
"Scamps!" Gramma called them. "Jackanapes!"
At first, they wouldn't grow at all. When they
did come up, they did NOT look healthy.

Gramma made them a special strengthening tonic.
They were listless.

She watered them with special water.
They got sicker.

She sprayed them with medicine.
They stayed puny and sad.

She covered them up at night with burlap blankets.
They stopped growing completely.

Then, one raw, rainy night, we had frozen peas with our dinner. Dad saw Gramma glaring at the ones on her plate.

"These are very special peas," he said. And he read from the box, " 'These lush Vegetables are harvested from Fertile Fields, frozen Farm Fresh and packed brim-full of Mother Nature's Goodness. With added salt.'

"What could be better?" he coaxed.

"Mother Nature can do better than *this!*" snapped Gramma.

And she folded up her napkin, jabbed it into its holder, and left the table.

We saw her staring out the window at the rain pelting down on her wispy bean plants.

We felt sorry for Gramma
and her poor beans.
 But Dad said, "Now don't
you fret about Gramma.
She wouldn't like that one
little bit! Let's scoot
on up and start your bath."

For the next two days,
Gramma kept herself busy
getting ready for the trip
she took every year. She
packed, cooked, cleaned,
and gave Dad big lists of
what to do while she was away.

Before she left, Gramma and I walked through the garden. We went past the pink roses, gold zinnias, yellow snapdragons, and purple pansies to the droopy, brown beans.

"Alright Dorothea," she said briskly, "I know you'll do your best."

It rained most of
the week she was gone.
The morning that
Gramma was due back,
we tried to do some of
the things on Dad's lists.
At the last minute,
we rushed out to pull up
the weeds.

But we just stood and stared.
We didn't hear the taxi doors slam.

"Well!" Gramma said
from behind us. *"Well!"*

In only a week, the beans had grown so much that they were starting to crowd the flowers. Gramma tried to make them stay on their poles, but they refused to behave themselves.

"Whippersnappers!" she scolded.
Dad teased her, "Are these
magic beans?"
"Certainly not," said Gramma.

But one day, we saw Gramma take some grapes without using the grape snips.

And one Saturday, we heard polka music coming from behind her closed door.

Finally, one night Dad tied our napkins into pirates' hats and we both sang "We gather together to ee-eat green be-eans…" as Gramma brought in a heaping bowlful.

She acted shocked and called us hooligans, but she hummed along with us as she filled our plates.

Gramma's beans still weren't as pretty
as frozen peas…
   …but I found that I couldn't stop
eating them!

"That's quite enough," Gramma said. "You are
not a starving barbarian. You will leave the pattern
on the plate, please."

The End

Library of Congress Cataloging-in-Publication Data

Thomas, Elizabeth, 1953-
Green beans / Elizabeth Thomas ; illustrated
by Vicki Jo Redenbaugh.
p.   cm.
Summary: Strict and proper Gramma, unhappy that her green
beans won't grow, leaves for a vacation and has a surprise when
she comes back.
ISBN 0-87614-708-2 (lib. bdg.)
[1. Beans—Fiction. 2. Gardening—Fiction. 3. Grandmothers—
Fiction.] I. Redenbaugh, Vicki Jo, ill. II. Title.
PZ7.T366627Gr   1992
[E]—dc20   91-27664
                                            CIP
                                            AC

Manufactured in the United States of America

2 3 4 5 6 7 8 9 10 01 00 99 98 97 96 95 94 93